Environmental Economics

Dr. Hemant Pathak

ISBN: 1484171462
ISBN-13: 978-1484171462

DEDICATION

Dedicated to Shri Sainath Maharaj the all omnipotent of world the most merciful.

CONTENTS

Foreword

Environmental Economics; provides a unique insight into the problems our planet faces in terms of clean environment, and what to do about it. This is the only books expressed comprehensive and interdisciplinary focus on Environmental Economics with the multidimensional approach.

This book made of 07 years consistently research on environmental issues, makes it ideal source for students, teachers, industrialist, environmental experts and economists.

This book provides an essential guide to researchers, it offers: various aspects of economics; on the challenges and experiences in present scenario.

Simply explained, Environmental Economics is an important book for all who wish to make a difference in how to plan and manage our Environmental resources.

Dr. Hemant Pathak

M.Sc. (Gold medalist), Ph. D.

Assistant Professor of Engineering Chemistry

Indira Gandhi Govt. Engineering College,

Sagar, MP, India

Glossary

Abatement : Reduction in Emissions

Abundance: A term that applies when individuals can obtain all the goods they want without cost. If a good is abundant, it is free.

Accelerator : The causal relationship between changes in consumption and changes in investment.

Acid Rain : The precipitation of dilute solutions of strong mineral acids, formed by the mixing in the atmosphere of various industrial pollutants -- primarily sulfur dioxide and nitrogen oxides with naturally occurring oxygen and water vapor.

Acquired Endowments: resources a country builds for itself, like a network of roads or an educated population

Assets: any item that is long-lived, purchased for the service it renders over its life and for what one will receive when one sells it What a person or business owns.

Average Costs : the total costs divided by the total output

Average Productivity : total quantity divided by the total quantity of input

Average Variable Costs : the total variable costs divided by the total output

Balance of Payments: A record of all the financial transactions between a country and the rest of the world during a given year.

Barriers To Entry: factors that prevent firms from entering a market, such as government rules or patents

Basic Competitive Model: the model of the economy that pulls together the assumptions of self-interested consumers, profit maximizing firms, and perfectly competitive markets

Benefit-Cost Analysis: A tally/comparison of expenditures and advantages in dollar terms resulting from various actions.

Benefits in Kind: Noncash forms of pay or assistance.

Capital: The existing stock of productive resources, such as machines and buildings, that have been produced.

Capital Intensive : Production methods with a high quantity of capital per worker.

Capital Gain: The increase in the value of an asset between the time it is purchased and the time it is sold

Capital Market: the market in which savings are made available to investors

Carbon Tax: a charge on fossil fuels (coal, oil, natural gas) based on their carbon content. When burned, the carbon in these fuels becomes carbon dioxide in the atmosphere, the chief greenhouse gas.

Climate Change: a regional change in temperature and weather patterns. Current science indicates a discernible link between climate change over the last century and human activity, specifically the burning of fossil fuels.

Cost : The most valuable opportunity forsaken when a choice is made.

Cost-Benefit Analysis: A tally/comparison of expenditures and advantages in dollar terms resulting from various actions.

Cost-Effectiveness Analysis: Least extensive way of achieving a given environmental quality target, or the way of achieving the greatest improvement in some environmental target for a given expenditure of resources.

Demand : The maximum quantities of some good that people will choose (or buy) at different prices. An identical definition is the relative value of the marginal unit of some good when different quantities of that good are available.

Developed Countries: the wealthiest nations in the world, including Western Europe, the United States, Canada, Japan, Australia, and New Zealand

Economic Growth : A sustained increase in total output or output per person for an economy over a long period of time.

Economic Regulations : The control of entry into the market, pricing, the extension of service by established firms and issues of quality control.

Economics : The study of choice and decision-making in a world with limited resources.

Equilibrium : The amount of output supplied equals the amount demanded. At equilibrium, the market has neither a tendency to rise nor fall but clears at the existing price.

Externalities: a situation in which an individual or firm takes an action but does not bear all the costs (negative externality) or receive all the benefits(positive externality) Costs or benefits that fall on third parties.

Good : Anything that anyone wants. All options or alternatives are goods. Goods can be tangible or intangible.

Indexation : Modifying contracts so that their dollar terms adjust to the inflation rate as measured in an index, such as the consumer price index.

Inflation : Increase in the overall level of prices over an extended period of time.

Interest : The annual earnings that are sacrificed when wealth is invested in a given asset or business. The interest sacrificed by investing in a given business is often called the cost of capital.

Intrinsic Values: Value that resides 'in' something and that is unrelated to human beings altogether.

Inventory : A stock of goods or resources held by a buyer or seller in order to reduce the cost of exchange or production.

Labor Productivity : The ratio of real output per unit of labor input; growth is measured by a higher ratio of outputs to inputs.

Law of Demand : People purchase more of any particular good or service as its relative price falls; they purchase less as its relative price rises.

Law of Supply : At higher relative prices, the quantity supplied of a good will increase; at lower relative prices, smaller quantities will be supplied.

Macroeconomics : The study of the sum total of economic activity, dealing with the issues of growth, inflation0 and unemployment and with national economic policies relating to these issues.

Marginal : The additional or extra quantity of something. If one drinks six sodas in a day, the marginal soda would be the sixth soda.

Marginal Cost : The increase in total costs as one more unit is produced.

Marginal Productivity : The additional output obtained by adding an additional unit of a productive resource, such as labor. More precisely, marginal productivity is the change in total output divided by the change in the amount of the productive resource employed.

marginal productivity = change in total output change in amount of productive resource

Market : A network in which buyers and sellers interact to exchange goods and services for money.

Market Clearing Price : A price which rations the supply of a good among competing consumers so that the quantity of the good demanded is equal to the quantity supplied.

Market Economy : A decentralized system where many buyers and sellers interact.

Microeconomics : The study of the individual parts of the economy, the household and the firm, how prices are determined and how prices determine the production, distribution and use of goods and services.

Minimum Wage : A wage below which employers may not legally pay employees for specific kinds of employment.

Net Worth : The difference between the assets and liabilities of a person or business.

Pollution Fee or Tax: Charge for the amount of waste or pollution. The charge makes it worthwhile for a producer to cut back, right up to the point where it begins to cost more to reduce pollution than to pay the tax.

Price :The amount of money, or other goods, that you have to give up to buy a good or service.

Profits : The excess of income over all costs, including the interest cost of the wealth invested. The net income of a business is not an accurate measure of its profit.

Public Goods : Goods that cannot be withheld from people even if they don't pay for them.

Quota : A quantitative restriction on imports.

Revenues : Total gross earnings of a firm before subtracting costs.

Tariff : A tax on imports.

Technological Change : An advance, usually scientific, that causes an increase in output to occur relative to the quantity of inputs.

Terms of Trade : The relative prices of goods and services traded in international markets.

Wealth : The value of the existing stock of goods; those goods may be tangible or intangible.

WTA (Willingness To Accept) : Minimum amount of money one would accept to forgo some good or to bear some harm.

Environmental Economics

1. Introduction

Environmental economics is a subfield of economics concerned with environmental issues. It is the study of environmental uses and abuses as viewed through the lens of economics. It is a distinct branch of economics that acknowledges the value of both the environment and economic activity and makes choices based on those values.

Economic concerns means market failure, externality, or valuation, are applied to environmental topics. Economics is divided into *microeconomics*, the study of the behaviour of individuals or small groups, and *macroeconomics*.

The study of the economic performance of economies as a whole. Environmental economics builds on the foundations of microeconomic analysis, a number of key features that make it an important field of study. The goal of environmental economics is to discover a balance between the least amount of usage and the greatest societal benefit.

Environment has become a scarce resource. Since economics is about how to deal with scarce resources, it can often be useful when tackling environmental problems.

Environmental policy may benefit the economy on the basis of following points like enhances Productivity, Stimulates Innovation, Increases Employment with quality point of view, Improves Balance of Trade, Strengthens Capital Base, Promotes Economic Cohesion, Encourages Transition to a Resilient, Supports Public Finances, and Sustainable Economy.

Environmental Economics undertakes theoretical or empirical studies of the economic effects of national or local environmental policies around the world.

2. Valuation in Environmental Economics

Environmental economics is the study of environmental problems with the perspective and analytical ideas of economics. Environmental economics draws from both sides, but primarily from microeconomics and Valuation measures human preferences for or against changes in the state of environments. It does not value the environment on its own.

Environmental Economics

Environmental economics, is concerned with the fundamental issue of allocating scarce resources among competing uses. One of the big issues in environmental economics is determining the value of natural resources.

Gold, have an actual monetary value assigned to it, and many things have a value based on their use and indirect use. However, it can be nearly impossible to determine the value of having a green Earth for future generations.

Economic and environmental objectives are often perceived as being contradictory. economic valuation is based on individual preferences and choices. People express their preferences through the choices and tradeoffs that they make, given certain constraints.

Other natural things, like value of an intact ozone layer or a lack of pollution, are intangible and therefore without a price tag. Many economists have been criticized for putting a 'price tag' on nature. While decisions are being made every minute regarding resource allocation.

An economist ensures the costs and benefits of environmental measures are well balanced. Environmental valuation can be a useful, yet also difficult and controversial tool. It is difficult to estimate costs and benefits. With the use of market-based instruments.

There are two types of values: use and non-use.

Use value may be defined as the value derived from the actual use of a good or service, such as hunting, bird-watching, or hiking etc.

Non-use values can be explained as passive use values, are values that are not associated with actual use, or even the option to use a good or service.

Non-use value is the most difficult type of value to estimate. Total economic value is the sum of all the relevant use and non-use values for a good or service.

The study of economics has always emphasized the relative scarcity of resources, whether they are natural, capital, or human, thereby placing constraints on what we can have and affecting the choices and decisions made by individuals or by society.

The costs and benefits of alternative environmental policies to deal with air pollution, water quality, toxic substances, solid waste, and global warming.

3. Importance of Cost/ Benefit Analyses in Environmental Economics

Economics is part of a science belongs to huminity contain certain theories, values, methods, and assumptions. While Environmental economics investigates and assesses different methods of reaching an efficient and equitable use of all resources from the society point of view, not just individual decision makers.

Cost Benefit Analysis (CBA) is used in project evaluation and regulatory review. The cost-benefit analysis is basically compiling the costs of a project as well as the benefits, then translating them into monetary terms and discounting them over time.

The use of natural resources and the deterioration of the environment is an expected part of life. Aim of economists is to understand how to produce goods for society in the most efficient manner. This is achieved by having a understanding of human activities in a global market. environmental benefits often lack market value, yet their costs are known.

Modernization and urbanization living relies more on energy and various types of nonrenewable resources, and consequently pollution are inevitable. This types expenditures are by no means good or healthy, they are a essential part of modern life and needed for the developing society.

An Environmental Economic analysis is not possible without understanding of how the project would affect the proposed area. When evaluation complete, the analyst should have a knowledge of proposed project impacts, classified according to the type of value they are likely to affect (use or non-use) and the group or groups that would benefit from the project.

Now scientists focus on reducing the carbon footprint, but environmental economists working in the field of cost benefit analyses.

In current scenario mild environmental ruin may be worth great economic benefit. this analysis must describe clearly what each benefit is and convert it into human needs.

4. Effect of Externality on Environmental Economics assessment

An externality exists when a person makes a choice that affects other people in a way that is not accounted for in the market price. For example a factory released air pollutant will typically not take into account the costs that its pollution imposes on surrounding living beings.

Heller and Starrett, defined externality as "a situation in which the private economy lacks sufficient incentives to create a potential market in some good and the nonexistence of this market results in losses of Pareto efficiency."

In economic terminology, the concept of externality refers to a cost or benefit incurred by someone other than the buyer or seller. This effect is generally not accounted for in the price the buyer paid. creation of externalities occurs when clear property rights are absent, as with air and some water resources.
Sometimes the government intervenes in an attempt to promote efficiency and bring the market back into equilibrium.

A positive externality provides a benefit; a negative externality results in a cost to society. In competitive markets, information exists about how much consumers value a particular good because we know how much they are willing to pay.

As natural resources are involved in the production of some goods other factors like scarcity issues, generation of pollution are not included in its production cost.

In these instances, scarcity issues or pollution become externalities, costs that are external to the market price of the product.

Externalities may include consideration issues like conservation and valuation of natural resources, pollution control, waste management and recycling, and the efficient creation of emission standards.

Environmental disasters can be devastating and widespread. When disasters strike on houses, industries or infrastructure whole development may be damaged or destroyed and people's livelihoods may be temporarily or permanently disrupted. Physical damage is the most visible economic impact. However, the long term impacts like lost income, through disruption of trade will be resultant.

5. Ecology economics

Environmentalists work on reducing the magnitude of pollution for improvement of society's ability to produce and advance. Economists work to ensure the best financial situation for the benefit of maximum number of people.

Environmental economists combine the two fields, encouraging economic and societal advancement without forgetting the effect they may be having on the environment.

Ecology economics is the study of the ecosystem and focuses on how natural resources, or the lack thereof, affect humans socially and financially.

Ecology economics is focused on protecting the environment. Many nations, the ability of the economy to meet basic needs allows them to focus more on environmental issues. While economists may believe that any increase in consumption is good, ecology economists question whether the consumption has improved quality of life.

6. Sustainability for development

Global economies use natural resources to sustain life. Increasing world population puts rising pressure on our natural wealth all the time. Many fear that our current path of production and population growth is not sustainable.

The United Nation's World Commission on Environment and Development commissioned a study on the subject by what is now known as the Brundtland Commission. The resulting report, Our Common Future (1987), defined sustainable development as "development that meets the needs of the present without compromising the ability of future generations to meet their own needs,".

This report also described three major factors to sustainable development:

a. Economic growth
b. Environmental protection
c. Social equity,

Also suggested that these three can be achieved by gradually changing the ways in which we develop and use technologies. sustainability is the long term potential for survival. current lifestyles are not sustainable because natural resources will eventually run out.

When people choose to ride a bike or walk, to recycle, or to install solar panels, they are making choices in line with sustainability.

The United Nations attempted to express these views in 1992 by their first Earth Summit in Rio de Janeiro. International community first time agreed on a comprehensive strategy to address development and environmental challenges through a global partnership.

The framework for this partnership was Agenda 21, which entirely covered the key aspects of sustainability – economic development, environmental protection, social justice, and democratic and effective governance.

7. Market economics

Two basic terms that are used by economists are demand and supply. how much of something people want known as the demand while How much of something that is provided is known as supply makes a working market.

Utilization of natural resources and whole environmental amenities in any ways. It is difficult to find the equilibrium through mere market pricing because they are not true market goods. Also difficult to analysis costs and benefits analysis of using or extracting natural resources while also taking into consideration future costs and benefits, as well as the intrinsic and existence value of the resources. When the market fails to allocate the resources efficiently, market failure can occur.

Market options can include economic incentives and disincentives, or the establishment of property rights. Environmental economics focuses on market failures, which are the rationale for considering government intervention.

Market failure means that markets fail to allocate resources efficiently. It occurs when the market does not allocate scarce resources to generate the greatest social welfare addresses the important opportunity costs of options in that the resources used for the environmental protection options could be used to yield other benefits.

8. Goods (Common/public)

Economics focuses on analysis at the margin of the actual choices that decision-makers face in selecting between the options to achieving greater levels of environmental improvements. it means that it will become increasingly more important to carry out careful economic analysis of the opportunity costs and trade-offs involved regarding the options, as the public demands greater environmental improvements.

When it is too costly to exclude some people from access to an environmental resource , the resource is called a public good. The mitigation of climate change effects is an example of a public good, where the social benefits are not reflected completely in the market price.

A country's incentive to invest in carbon abatement is reduced because it can "free ride" off the efforts of other countries. Assessing the economic value of the environment is a major topic within the field. Use and indirect use are tangible benefits accruing from natural resources or ecosystem services

These challenges have long been recognized. Commons refers to the environmental asset itself, "common property resource" or "common pool resource" refers to a property right regime that allows for some collective body to devise schemes to exclude others, thereby allowing the capture of future benefit streams; and "open-access" implies no ownership in the sense that property everyone owns nobody owns.

9. Effect of Environmental regulations on economics

An environment impact on any nations has to be estimated by the Government agency is done using cost-benefit analysis. If pollution rises above the threshold limit environmental regulations are enforced by fines in the form of tax.

For a developing nations it is must of healthy populations, for getting this goal pollution must be regularly monitored and strict laws must be enforced.

Command and control regulation applies to determined emissions limits on pollutants, even though each industries has different costs for emissions reductions. In Some industry control on pollution inexpensively, while others can abate at high cost. Because of this, the total abatement has expensive and need of efforts is must to control on pollution. This is the challenge for a Environmental economist to find the cheapest emission abatement efforts. Many industries are enforced to obey these regulation as a viable way to promote their greening practices in a globalizing economy.

10. Green economy

Green economy is an economy or economic development model based on sustainable development and a knowledge of ecological economics. It is an indistinct form of economy is considered to be component of the ecosystem in which it resides. It is viewed as more pragmatic in a current price system. regulations and conservation.

Objective of green economy is to improved human well-being and social equity, while significantly reducing environmental risks due to erosion, water quality problems, diseases, desertification, and other outcomes and ecological scarcities may cause loss of natural capital.

For an overview of the developments in international environment policy that led up to the UNEP Green Economy Report.

Pollution reductions should be achieved by way of tradeable emissions permits, which if freely traded may ensure that reductions in pollution are achieved at least cost, then a firm would reduce its own pollution load only if doing so would cost less than paying someone else to make the same reduction. In practice, tradeable permits approaches have had some success, such as the U.S.'s sulphur dioxide trading program or the EU Emissions Trading Scheme, and interest in its application is spreading to other environmental problems.

A pollution tax that reduces pollution to the socially optimal level would be set at such a level that pollution occurs only if the benefits to socicty exceeds the costs. Some advocate a major shift from taxation from income and sales taxes to tax on pollution like green tax shift.

11. Green accounting, trading and Finance

Environmental economics was a major influence for the theories of natural capitalism and environmental finance concerned with resource conservation in production, and the value of biodiversity to humans.

Green accounting is a type of accounting that attempts to factor environmental costs into the financial results of operations. . It is an assessment and evaluation of the results, costs, and savings attributable to environmental protection activities. It has been seemed that gross domestic product ignores the environment and therefore decision makers need a revised model that incorporates green accounting.

Green trading encompasses all forms of environmental financial trading, including carbon dioxide, sulfur dioxide, nitrogen oxide, renewable energy credits, and energy efficiency. It accelerate change to a cleaner environment by using market-based incentives whose application is global.

All these emerging and established environmental financial markets have one thing in common, which is making the environment cleaner by either reducing emissions, using clean technology or not using energy through the use of financial markets.

Environmental Finance is the use of various financial instruments to protect the environment. The field is part of both environmental economics and the conservation movement.

An increasing commitment by the UNEP to the ideas of natural capital and full cost accounting under the banner 'green economy' could blur distinctions between the schools and redefine them all as variations of "green economics". responsible for global monetary policy have stated a clear intention to move towards biodiversity valuation and a more official and universal biodiversity finance. Taking these into account targeting not less but radically zero emission and waste is what is promoted by the Zero Emissions Research and Initiatives.

12. Application of Environmental Economics

Natural resources are a precious commodity on earth. global citizens should be aware to the earth environmental problems and must do serious effort to preserve earth, like recycling or buying reusable materials. Environmental affect on the market must be studied and analyzed.

The eco-industry is an important source of new jobs, it is a business, investment, and technology-development model that employs market-based solutions to balancing the world's energy needs and environmental integrity.

An ecological-economical model provides a means to account for and value land management activities that improves the condition of natural capital and values the output of eco-services, it provides the framework to build an ecological intelligence system that allows the public arena of commerce to express sustainability.

Green technology are advance science to funding generated through the voluntary carbon offset market in the emerging word.

Today every industries are seeking ways to clean up their environmental impact. Bad energy practices that they cannot eliminate, they may offset. Modern labs actively engaged in developing cleaner energy practices and increasing energy efficiency for the future.

Use of green trading and green finance and eco-commerce allows for the further development of clean technologies such as wind power, solar power, biomass, and hydropower. A good environmental policy contributes to a employment towards jobs associated with cleaner, more efficient products and processes.

Now It can say that environmental economics is an important science being used more often for discussing environmental issues. Whether utilized as a medium to acertain which projects have the greatest beneficial or to determine natural resource damages, those individuals who have an understanding of some of the concepts, will have a extra advantage.

13. References

Arrow, K.J. (1951). *Social Choice and Individual Values*. New York: Wiley.

Baldwin, R. and Veljanovski, C.C. (1984). Regulation by cost-benefit analysis. *Public Administration* 62, 51-69.

I.J. Bateman, R.T. Carson, B. Day, M. Hanemann, N. Hanley, T. Hett, M. Jones-Lee, G. Loomes, S. Mourato, E. Ozdemiroglu, D.W. Pearce, R. Sugden and J. Swanson (2002) Economic Valuation With Stated Preference Techniques: A Manual, Edward Elgar Publishing, Cheltenham.

Becker, G.S. (1993). Nobel lecture: the economic way of looking at behaviour. *Journal of Political Economy* 101(3), 385-409.

Beckerman, W. and Pasek, J. (1997). Plural values and environmental valuation. *Environmental Values* **6**, 65-86.

Bishop, M., Kay, J. and Mayer, C. (1995). Introduction. In Bishop, M., Kay, J. and Mayer, C. (eds), *The Regulatory Challenge*. Oxford: Oxford University Press.

Breyer, S. (1982). *Regulations and its Reform*. Cambridge, Massachusetts: Harvard University Press.

The U.N. Department of Economic and Social Affairs, Division of Sustainable Development , Agenda 21. www.un.org/esa/sustdev/documents/agenda21/english/agenda21toc.htm

International Institute for Sustainable Development, www.iisd.org

Environment Agency (2000). *Risks and Values Framework*. Bristol: Environment Agency.

Foster, J. (ed.) (1997). *Valuing Nature? Economics, Ethics and the Environment*. London: Routledge.

Harsanyi, J. (1955). Cardinal welfare, individualistic ethics and interpersonal comparisons of utility. *Journal of Political Economy* 63: 309-321.

Helm, D. (1998). The assessment: environmental policy – objectives, instruments, and institutions. *Oxford Review of Economic Policy* 14(4), 1-19.

Lave, L.B. (1996). Benefit-cost analysis: do the benefits exceed the costs? In Hahn, R.W. (ed.), *Risks, Costs and Lives Saved: Getting Better Results from Regulation*. New York: Oxford University Press.

Layard, R. and Glaister, S. (eds) (1994). *Cost-Benefit Analysis*. Second edition. Cambridge: Cambridge University Press.

Mishan, E.J. (1988). *Cost-Benefit Analysis*. Fourth edition. London: Routledge.

Misser, H.J. and Quade, E.S. (eds) (1985). *Handbook of Systems Analysis: Overview of Uses, Procedures, Applications and Practice*. New York: Elsevier Science Publishing.

Nagel, S. (1984). *Public Policy: Goals, Means, and Methods*. New York: St. Martins Press.

Pearce, D. (1998). Cost-benefit analysis and environmental policy. *Oxford Review of Economic Policy* 14(4), 84-100.

Pearce, D. (2000). Controversies in environmental valuation. In McMahon, P. and Moran, D. (eds), *Economic Valuation of Water Resources: Policy and Practice*. London: Chartered Institution of Water and Environmental Management.

Pearce, D. and Barbier, E.B. (2000). *Blueprint for a Sustainable Economy*. London: Earthscan.

Pearce, D.W. and Nash, C.A. (1981). *The Social Appraisal of Projects: A Text in Cost-Benefit Analysis*. Basingstoke: Macmillan.

Porter, M.E. and Van der Linde, C. (1995). Toward a new conception of the environment-competitiveness relationship. *Journal of Economic Perspectives* **9**, 97-118.

Robbins, L. (1932). *An Essay on the Nature and Significance of Economic Science*. London: Macmillan.

Sagoff, M. (1988). *The Economy of the Earth*. Cambridge: Cambridge University Press.

Wikipedia: the free encyclopedia

ABOUT THE AUTHOR

Dr. Hemant Pathak held positions as Assistant Professor in the department of chemistry, Govt. Indira Gandhi Engineering College, Sagar, MP, India. He had extensive experience in teaching, research and administrative management.

Dr. Pathak received his Ph.D. degree in chemistry from Dr. Hari Singh Gour Central University, Sagar, India and M.Sc. Gold medalist from Jiwaji University, Gwalior. He has published 08 books and more than 50 research papers in reputed International and National journals and received several awards. He is a member of editorial boards and reviewer boards of several international journals and societies. His area of specialization includes Engineering Chemistry and Environmental Pollution management.

www.ingramcontent.com/pod-product-compliance
Lightning Source LLC
Chambersburg PA
CBHW081420170526
45166CB00010B/3419